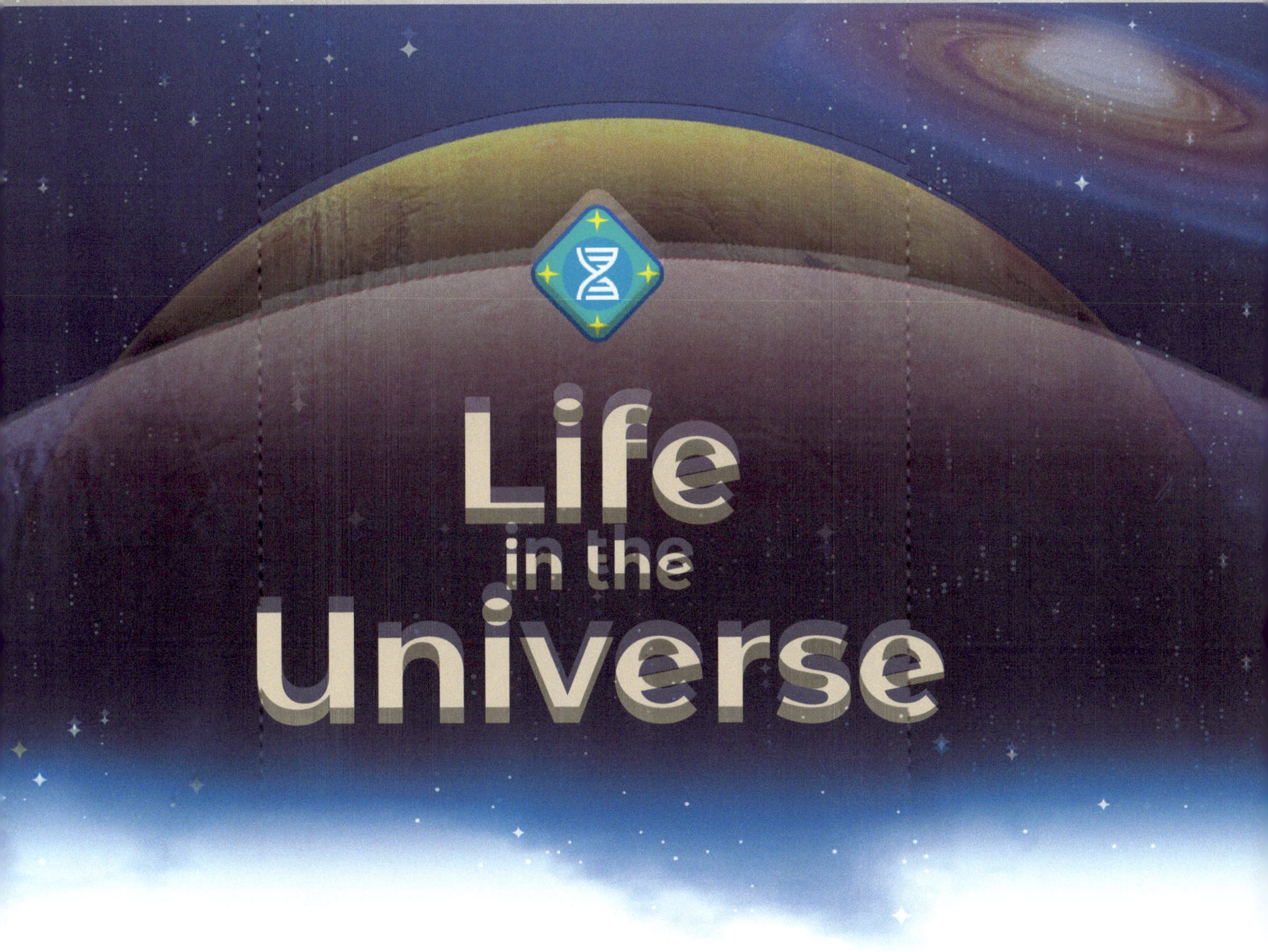

Life in the Universe

written and illustrated by Bernard Isselin

100 billion planets to discover

Are we alone in the universe?

We live on Earth, one of the planets* in the Solar system*. The entire system, planets and sun, orbits within an enormous vortex of billions of stars called a galaxy*. These stars, when viewed from Earth, create a diffuse milky band in the night sky, the Milky Way. Since all of these stars also have planets, one may wonder if life exists on any of them. To answer this question, it would be best to go and see for ourselves. However, our current spacecraft are too slow to travel the enormous distances that separate us from these other worlds. Therefore, the search for life in space is currently limited to the exploration of the planets in our Solar system by unmanned probes controlled from Earth.

HERE IS OUR GALAXY
It contains between 200 and 400 billion stars, and probably over 100 billion planets. Astronomers estimate that there are 2,000,000,000,000 galaxies in our Universe.

Leaving our planet

Can we live in interplanetary space if we leave our planet?

In space, there is no air to breathe or water to live on. It is a vacuum. Astronauts need a spacesuit filled with Earth's air to breathe and restore atmospheric pressure. This suit must also protect these brave explorers from the unbearable temperature differences that exist in space, depending on whether they are in shade or sunlight. Finally, the spacesuit is made of a thick, impermeable fabric that protects against invisible and dangerous cosmic rays coming from the Sun and other stars. If these rays were to directly hit the molecules that make up our bodies, they could cause burns and significant illnesses. Therefore, life in space is only possible inside a spacesuit or spacecraft.

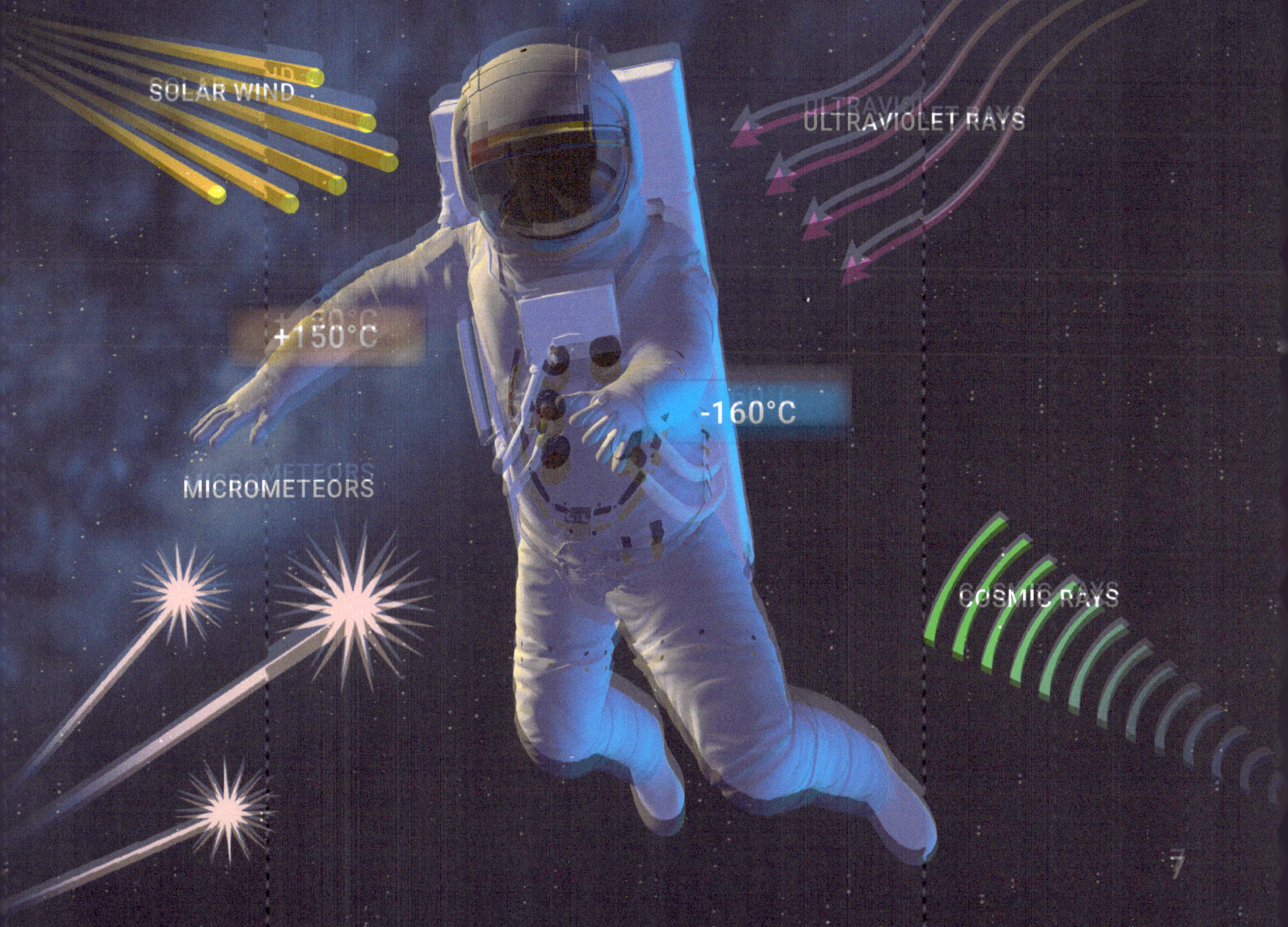

SOLAR WIND

ULTRAVIOLET RAYS

+150°C

−160°C

MICROMETEORS

COSMIC RAYS

All different, one chemistry

How does life on Earth work?

The inhabitants of Earth, including microbes, algae, fungi, plants, and animals, are made up of elements that are present on the planet. These elements are primarily the atoms of carbon, hydrogen, oxygen, and nitrogen, which are assembled into various molecules that interact with each other in liquid water. Perhaps a form of life based on other elements exists somewhere? But we do not know it. We cannot imagine what it would be made of or what it would look like. For now, and for these reasons, searching for extraterrestrial life means exploring worlds on which our four elements and liquid water also exist.

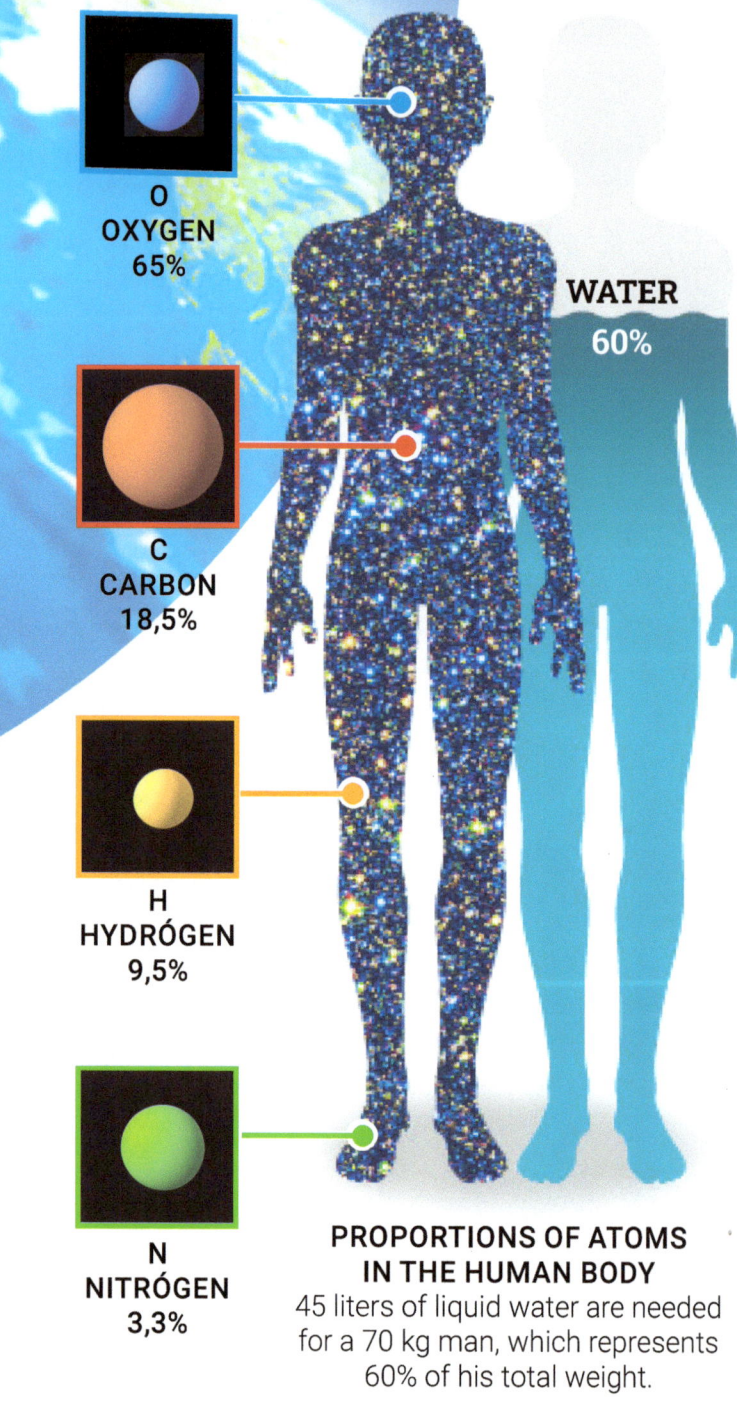

O
OXYGEN
65%

C
CARBON
18,5%

H
HYDRÓGEN
9,5%

N
NITRÓGEN
3,3%

WATER
60%

PROPORTIONS OF ATOMS IN THE HUMAN BODY
45 liters of liquid water are needed for a 70 kg man, which represents 60% of his total weight.

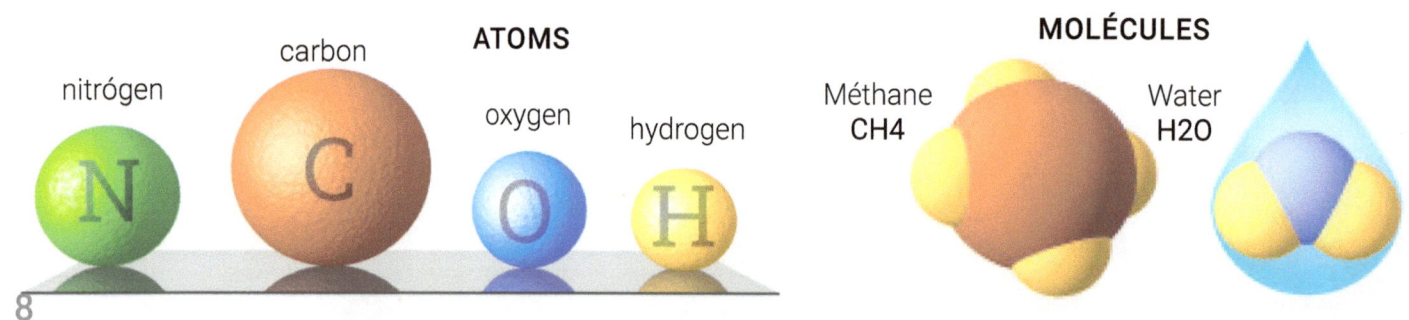

ATOMS — nitrógen, carbon, oxygen, hydrogen

MOLÉCULES — Méthane CH4, Water H2O

A problem of temperatures and pressures

SUN 6000°
Mercury 140°
Vénus 470°

TOO HOT

Are there other plants or animals on other planets?

Plants, animals, and humans are made up of 60% liquid water. If water freezes, life is petrified; if water boils, living matter is destroyed. Around the Sun, there is a zone called the «habitable zone,» in which it is neither too hot nor too cold. Earth is located exactly in the center of the habitable zone: It is a perfect habitable planet. Venus is too close to the Sun, Mars too far away, and the temperature and pressure conditions on their surface do not allow for the presence of liquid water. The many automatic devices sent from Earth to the surface of these two planets have found no evidence of life.

AT THE LIMITS OF TOLERANCE

Most living beings adapt to the average temperature on Earth, which is 15°C. Tiny «extremophiles» choose boiling or freezing water, acid, salt-saturated, radioactive water from nuclear plants, or the water of oceans 11,000 meters deep.

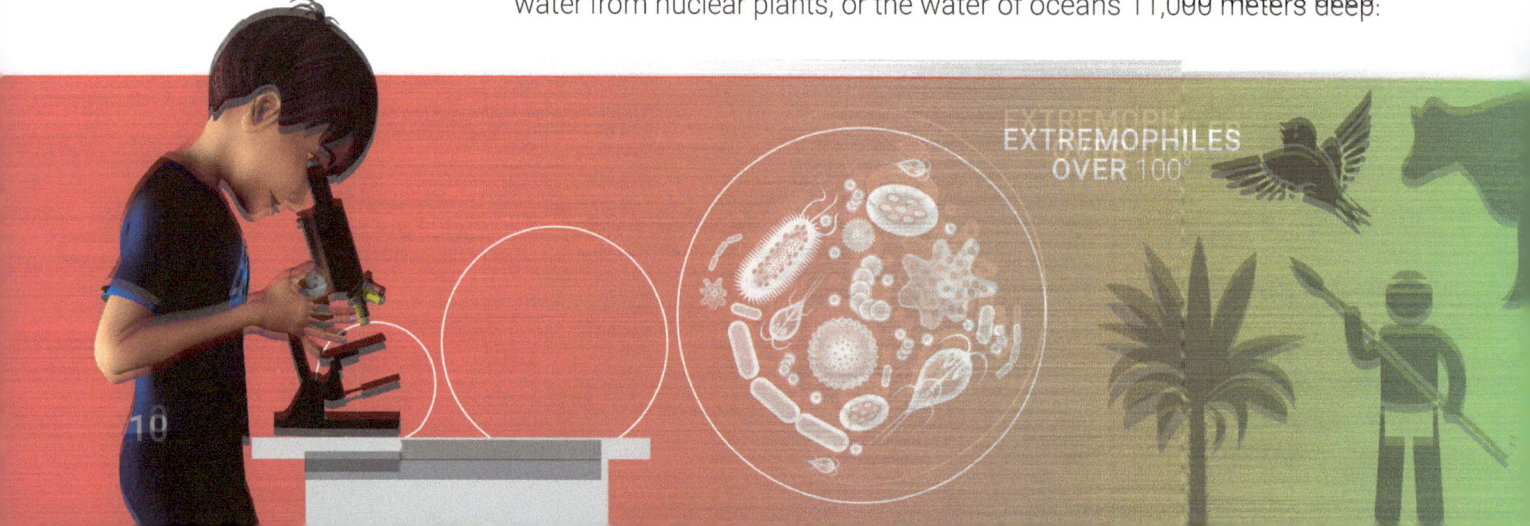

EXTREMOPHILES OVER 100°

Earth and Moon 15°

Mars -50°

HABITABLE ZONE

TOO COLD

VENUS

90 kg/cm²

300°

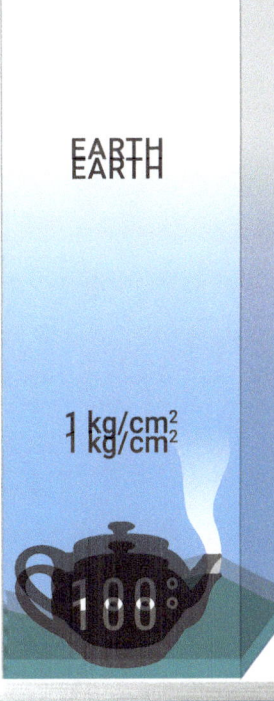

EARTH

1 kg/cm²

100°

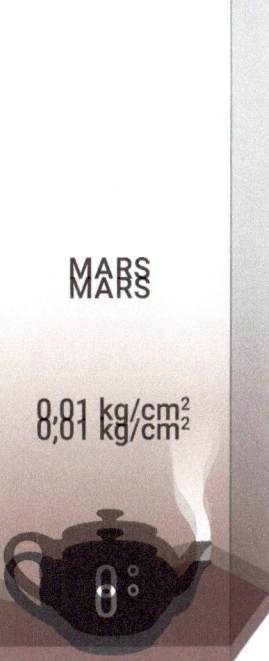

MARS

0,01 kg/cm²

0°

WHEN THE ATMOSPHERE PUTS PRESSURE

The layer of air that envelops the Earth weighs 1 kilogram per square centimeter at sea level. This pressure regulates the boiling temperature of water to 100°C. On Venus, the pressure reaches 90 kilograms per square centimeter, and water only begins to boil at 300°C. The atmosphere on Mars weighs 100 times less than on Earth, and water boils at 0°C.

EXTREMOPHILES
BELOW 0°

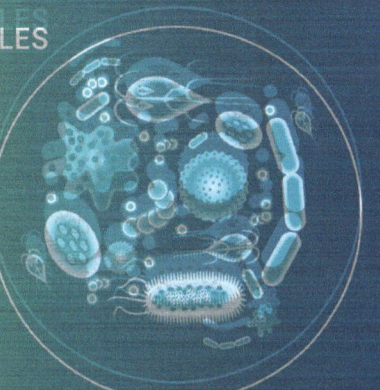

On the trail of water

In the UNIVERSE, out of 1,000,000 atoms, the most numerous are hydrogen and oxygen, that is to say, the atoms that make water.

HYDRÓGEN	910 580
OXYGEN	800
CARBON	300
NITRÓGEN	100
OTHER ATOMS...	88 220

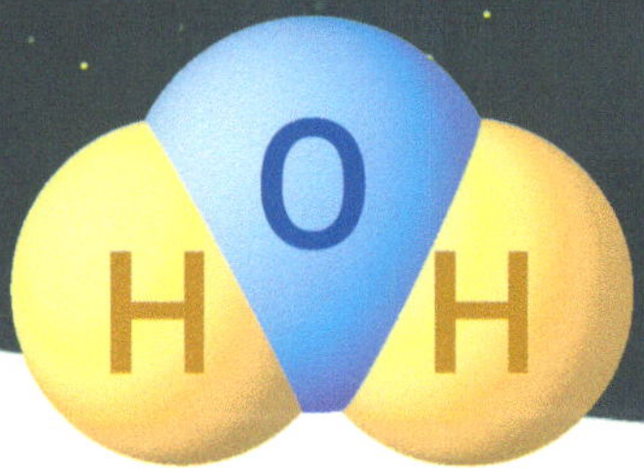

These atoms are produced by stars and since there are many stars in the galaxy, there are also many atoms.

Is there a lot of water in the universe?

Space probes and radio telescopes have found water (H2O) in the form of vapor near the Sun, or mixed with other gases in interstellar clouds. Some moons of giant planets are made up of half water ice and half rocks. Finally, the other atoms of life, carbon (C) and nitrogen (N), are also very abundant. But it is liquid water that life needs, and so far we only know of one planet that has it: Earth.

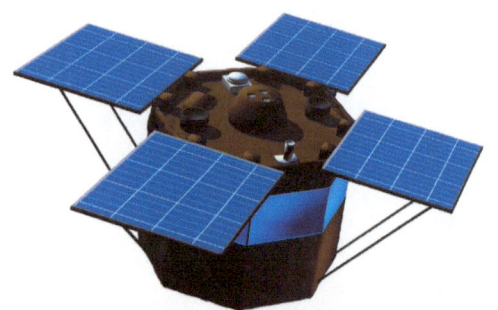

WATER IN THE SOLAR SYSTEM
Instruments aboard space probes and radio telescopes have revealed the presence of water in space and on many celestial bodies.

Where to find water?

COMETS AND METEORITES contain more than 80% water and some are very rich in carbon.

THE PLANETARY RINGS of Saturn contain 99.9% of frozen water in the bright regions. The dark stripes are made of carbonaceous matter.

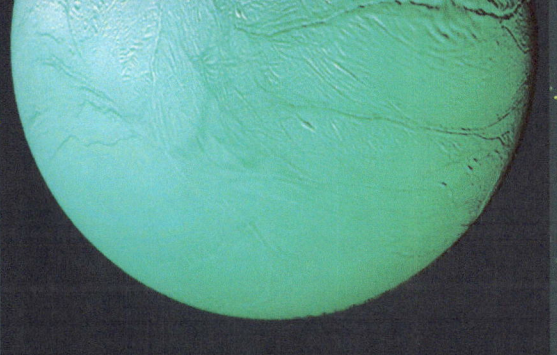

SOME MOONS OF GIANT PLANETS are made up of 50% rocks and 50% water in the form of an ocean covered by ice.

IF THE POLAR CAPS of the planet Mars melted, the ground would be covered with a layer of water over 10 meters deep.

ON THE MOON, THERE'S NO AIR, NO LIQUID WATER, BUT «FOSSIL» ICE!

Sometimes, comets covered with frozen water crash into the moon. If a large amount of this ice is vaporized and ejected far into space by the impact, some can fall back to the ground and reconstitute as ice at the bottom of some craters perpetually in the shadow.

COMET

FROZEN WATER

Dimension, Gravity, Water and Atmosphere

Why is there no water on the moon?

A long time ago, as a result of a collision with another planet, a piece of Earth was ejected. It started to orbit around it and became our satellite, the moon. So, the moon is a piece of Earth. Like Earth, the moon is also in the habitable zone. But the moon, being smaller, lacks an atmosphere, and without an atmosphere to exert pressure, there can be no liquid water. Without liquid water, life could not develop. However, the ice in the craters will be used to provide drinking water for the occupants of future permanent lunar stations.

MOON
8.3 kg
6 meters

EARTH
50 kg
1 meter

TO RETAIN AN ATMOSPHERE AROUND IT, A CELESTIAL BODY MUST HAVE CERTAIN DIMENSIONS:

A child weighing 50 kg who jumps 1 m high on Earth would weigh only 8.3 kg on the moon and would jump 6 m. This is because the moon is 6 times smaller than Earth, and its gravity, the force that attracts towards the ground, is 6 times weaker. Air is made up of molecules that «bounce» in all directions. On Earth, some move away into the sky, but gravity always pulls them back to the ground. On the moon, the attraction is insufficient to retain the molecules, and they have escaped into space long ago. There is no longer an atmosphere on the moon.

On the fleeing planet

Are there Martians?

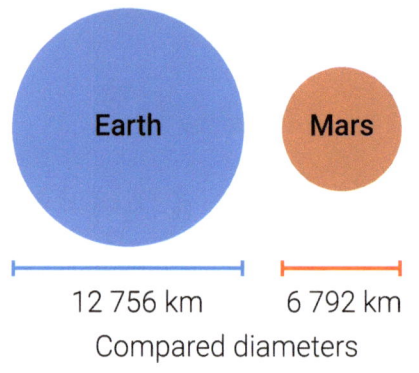
Earth — 12 756 km
Mars — 6 792 km
Compared diameters

The planet Mars, which is small and has a weak gravity, sees its atmosphere slowly escaping into space. The pressure at the surface is one hundredth that of Earth, while the average temperature on the planet drops to -60°C. Under these conditions, liquid water cannot exist. Yet the surface of Mars shows many riverbeds. Perhaps in the past, when atmospheric pressure was higher, water flowed on Mars. Life may have appeared in the depths of some oceans? Today, there remains the frozen water of the polar caps* and perhaps a little liquid water in the planet's subsoil, where the pressure is greater. Since 1974, many robots have landed on Mars in search of any trace of past or present life.

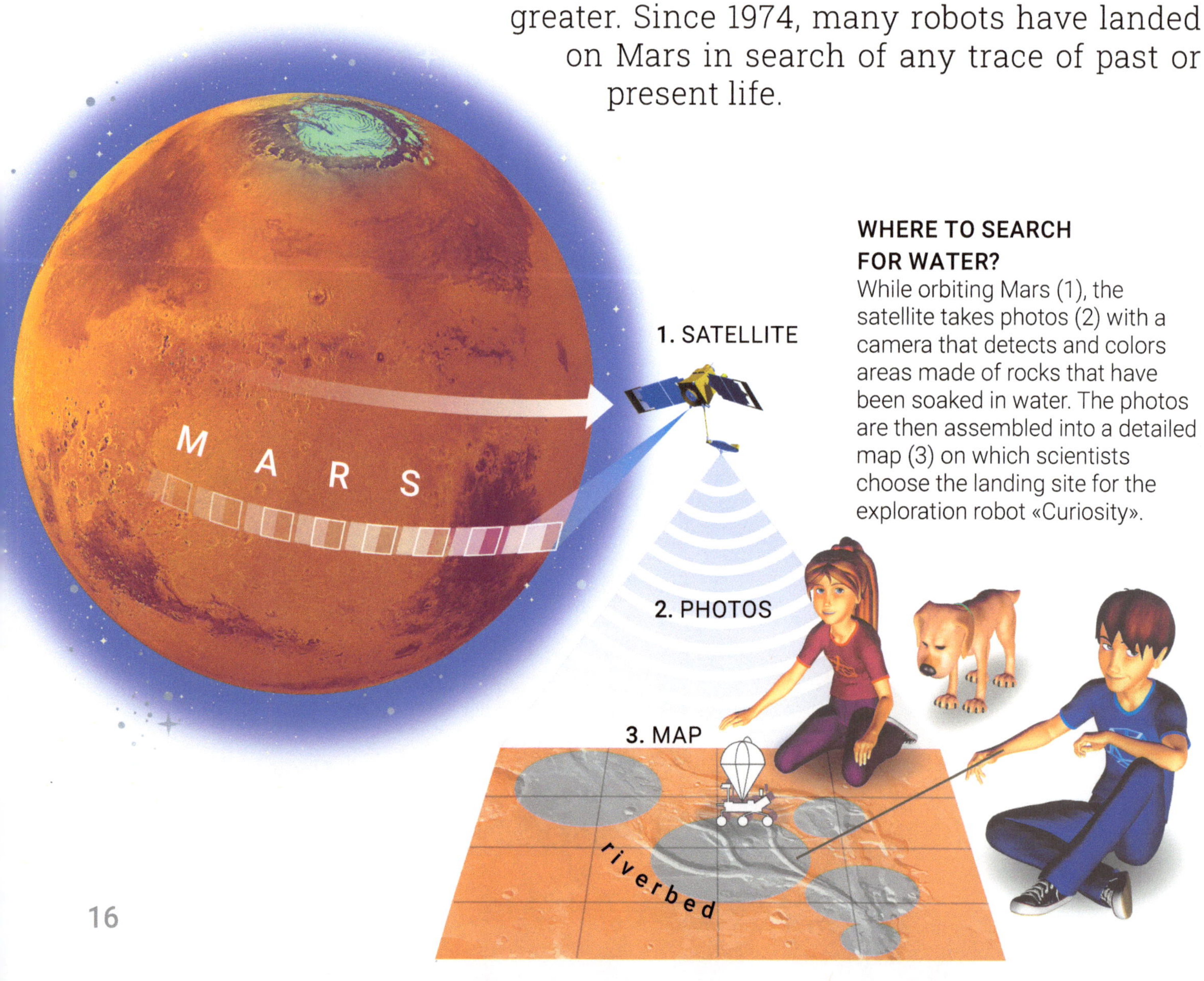

WHERE TO SEARCH FOR WATER?
While orbiting Mars (1), the satellite takes photos (2) with a camera that detects and colors areas made of rocks that have been soaked in water. The photos are then assembled into a detailed map (3) on which scientists choose the landing site for the exploration robot «Curiosity».

1. SATELLITE
2. PHOTOS
3. MAP
riverbed

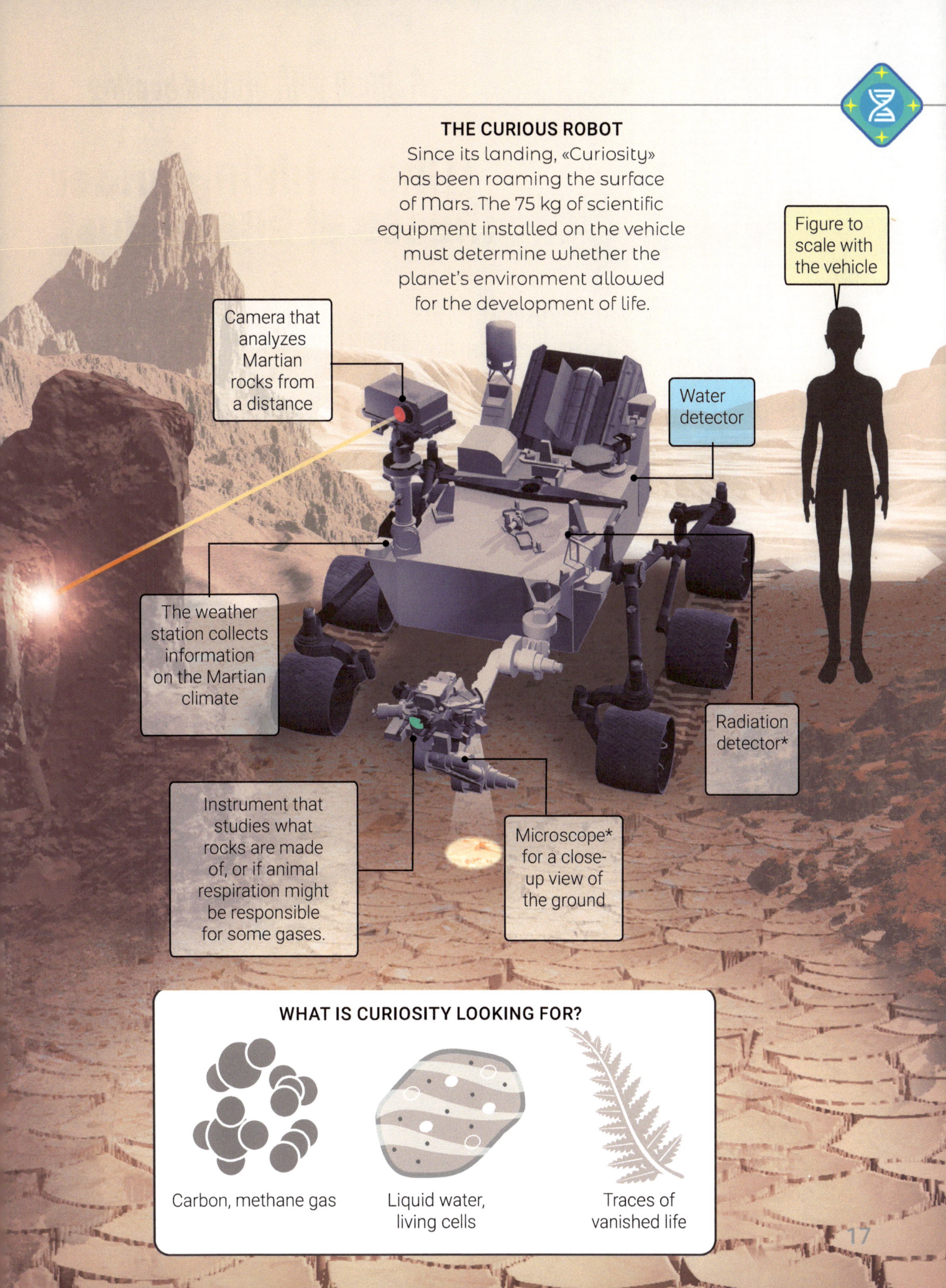

A moon with ground heating

Who is hiding under the ice of Enceladus?

Beyond Mars, the zone of the giant planets begins. These are Jupiter, Saturn, Uranus, and Neptune, all enveloped in poisonous gas cocoons. Totally uninhabitable, these celestial bodies are also surrounded by many moons (166 in all), some of which are very interesting. For example, Enceladus, a moon of Saturn, is made up of half a globe of rocks and half an immense ocean of liquid water covered by an ice sheet*. How can the liquid water that forms this ocean exist, when outside there is a freezing cold at –200°C? The phenomenon comes from the tidal forces* exerted by Saturn and the other moons. These forces knead Enceladus' rocky core, and the rubbing of the rocks against each other produces enough heat to melt some of the ice.

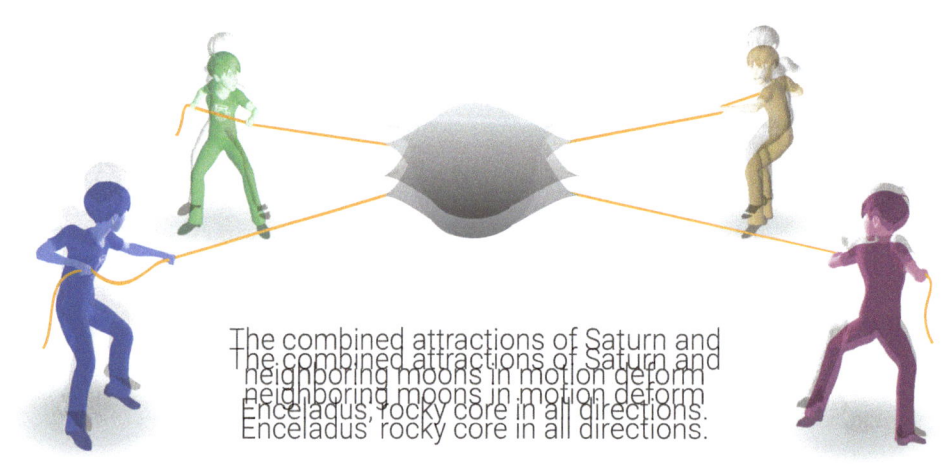

The combined attractions of Saturn and neighboring moons in motion deform Enceladus' rocky core in all directions.

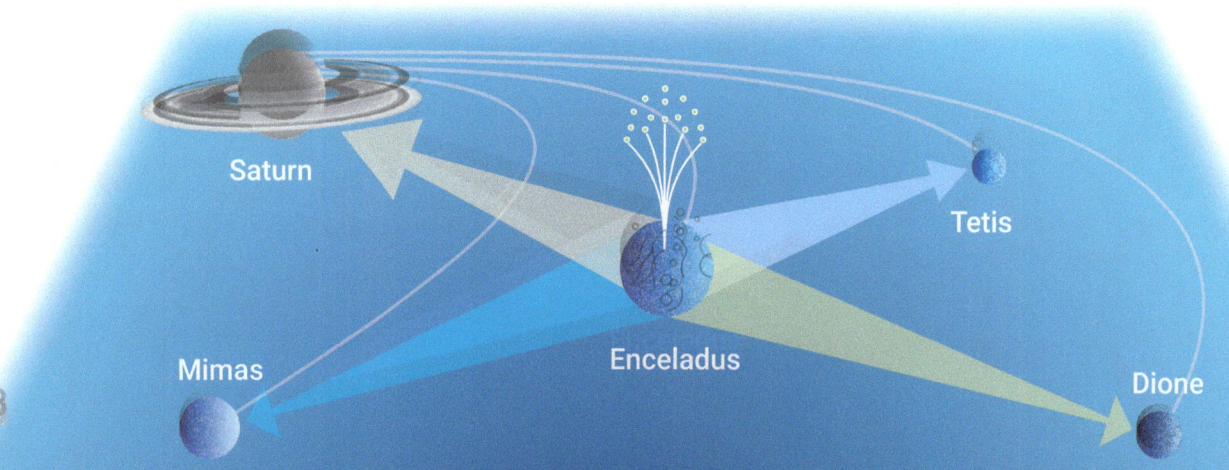

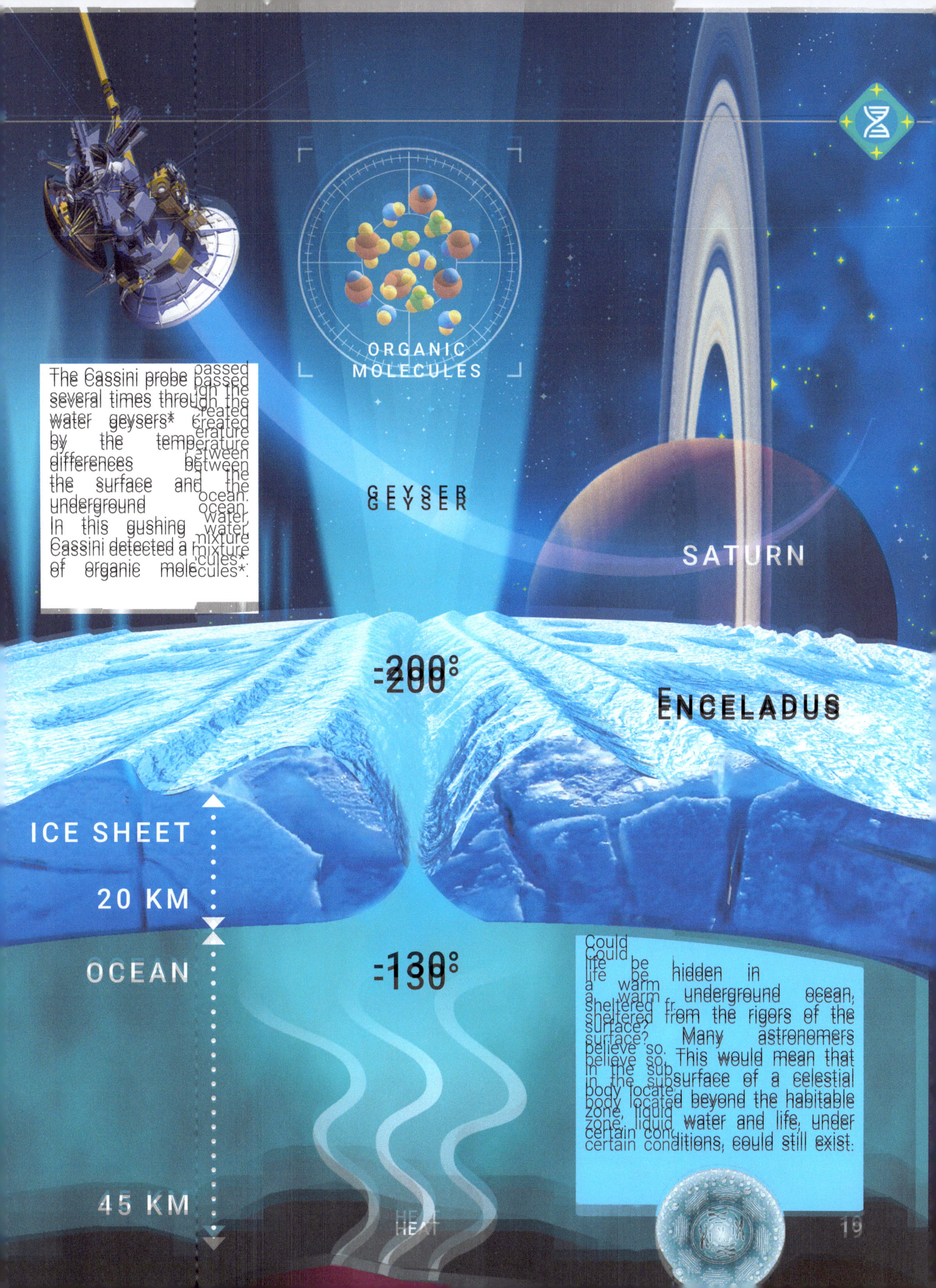

ORGANIC MOLECULES

The Cassini probe passed several times through the water geysers* created by the temperature differences between the surface and the underground ocean. In this gushing water Cassini detected a mixture of organic molecules*.

GEYSER

SATURN

-200°

ENCELADUS

ICE SHEET

20 KM

OCEAN

-130°

Could life be hidden in a warm underground ocean, sheltered from the rigors of the surface? Many astronomers believe so. This would mean that in the subsurface of a celestial body located beyond the habitable zone, liquid water and life, under certain conditions, could still exist.

45 KM

HEAT

Life seeks a planet that orbits smoothly

Are other stars surrounded by habitable planets?

The habitable zone around an average star like the Sun will also be located at the same distance as Earth, about 150 million kilometers away. However, there are stars, hotter or colder, smaller, «dwarfs,» or larger, «giants,» which means that the habitable zone will be farther or closer to the star. And then the orbits of the planets in the solar system are almost all circular. The distance between them and the Sun varies very little. The solar system seems very «wise and orderly,» with a habitable zone that stays in place. In contrast, many exoplanets orbit along very elongated orbits. During their years, the distance between them and their stars constantly changes.

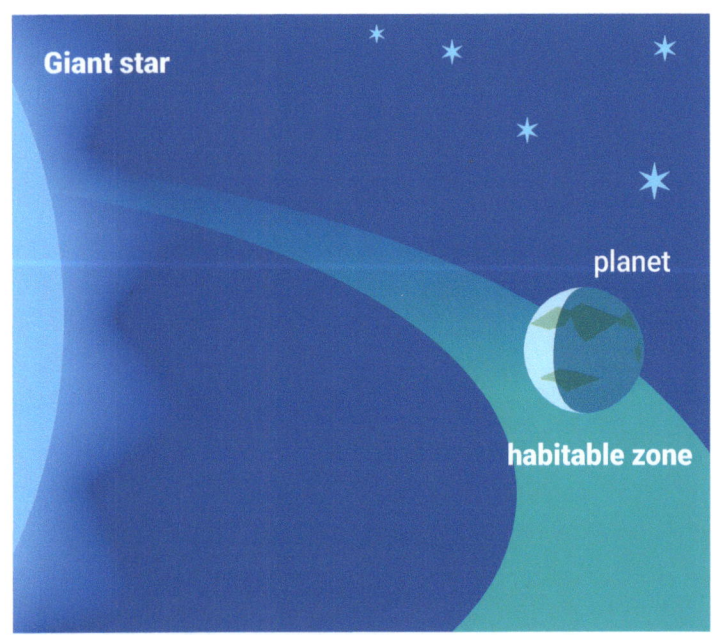
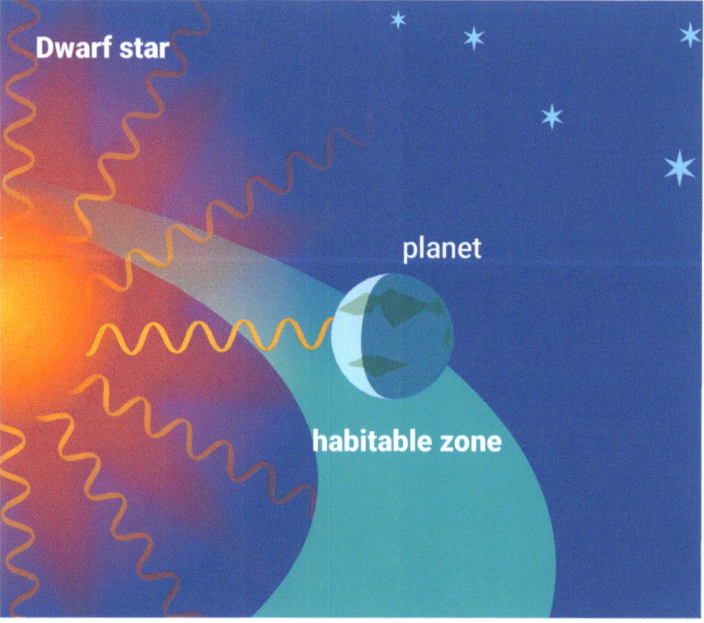

THE LIFESPAN OF STARS AND LIFE ON PLANETS: A giant star emits a lot of heat and the habitable zone is far away. However, a giant star burns too quickly for life to appear on a planet. Conversely, a dwarf star, which is less hot, burns slowly, but since the habitable zone is closer, it is subject to a continuous bombardment of particles and deadly rays emitted by very frequent eruptions in small stars.

NOT TOO HOT AND NOT TOO COLD
Life can adapt to changes in temperature, but it takes a long time. A planet with a very elongated orbit only passes through the habitable zone and goes from scorching to freezing several times. No living organism can develop under such conditions.

The colors of life on the skies' lands

Is there life on exo-earths?

Like Earth, exo-earths are worlds made mostly of rock and a metal core. If they are located at a good distance from their star and have an atmosphere, we can imagine that they are also habitable for living beings. But these planets are currently out of reach of our spacecraft. For now, we know these distant worlds by the light we receive when they pass in front of their star. The rays of light from the star near the planet can reveal the existence of an atmosphere around it and inform us about the gases that make it up. It is also important to know that a molecule of oxygen, composed of 2 atoms (O_2), only exists if a living being creates it through respiration. Thus, any trace of O_2 in the atmosphere of an exo-earth would signal the presence of life on that distant world. Other gases, such as methane (CH_4), can also be the result of animal life.

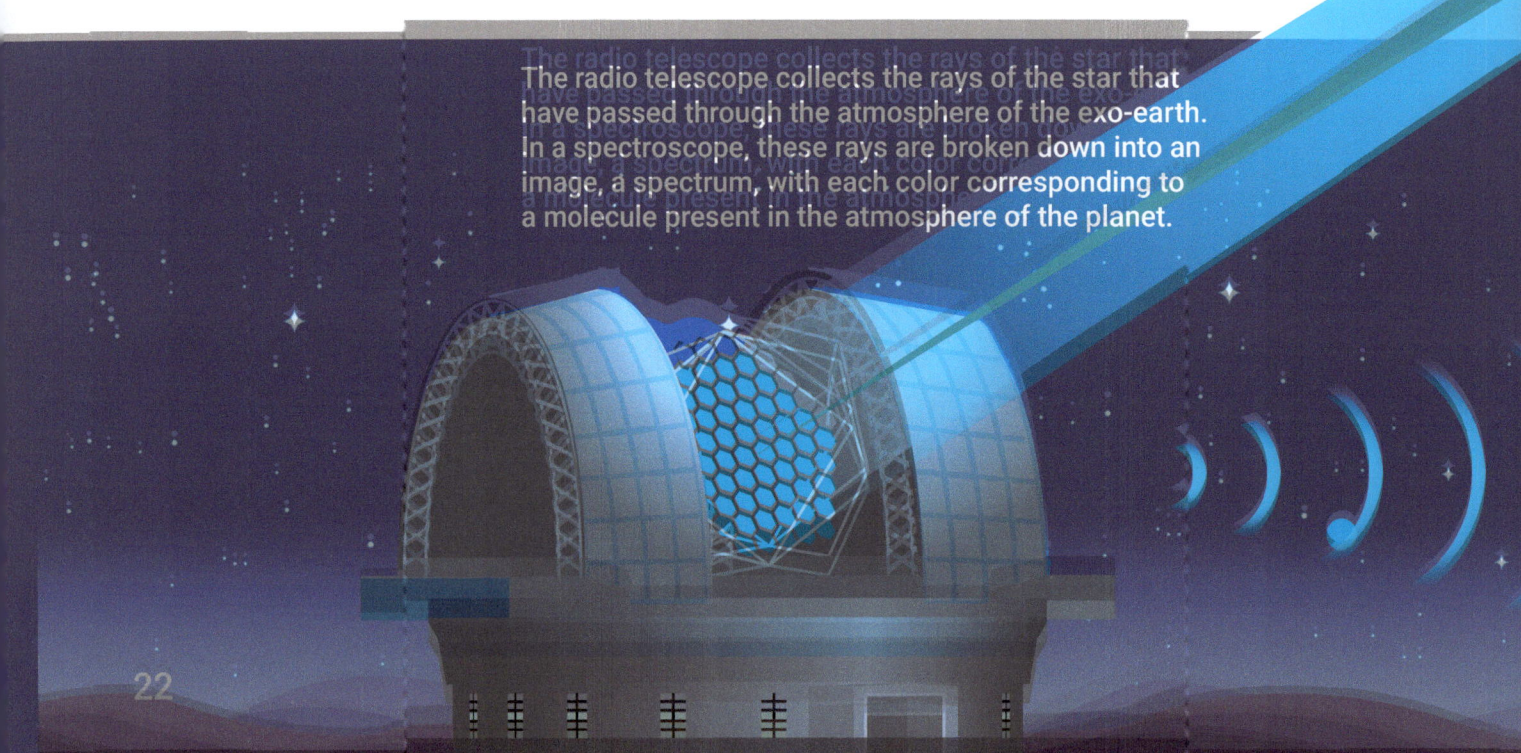

The radio telescope collects the rays of the star that have passed through the atmosphere of the exo-earth. In a spectroscope, these rays are broken down into an image, a spectrum, with each color corresponding to a molecule present in the atmosphere of the planet.

STAR

atmosphere

EXO-EARTH

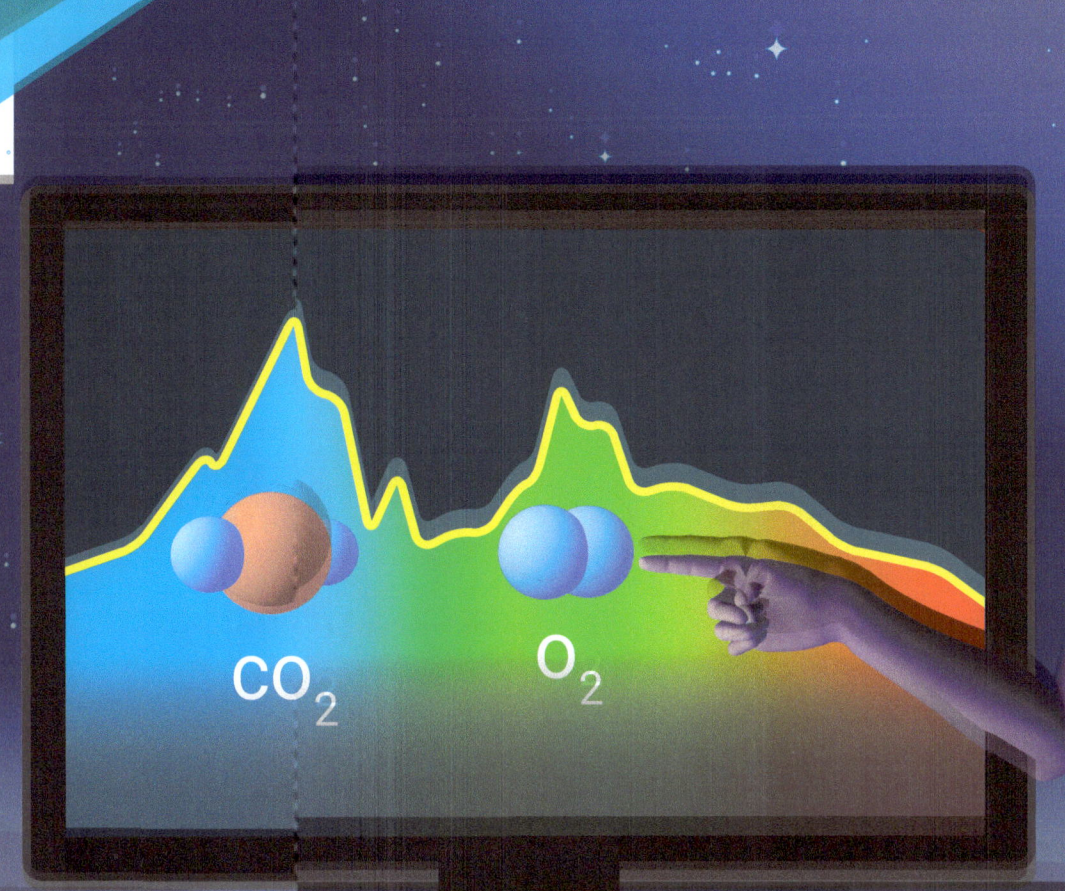

CO₂   O₂

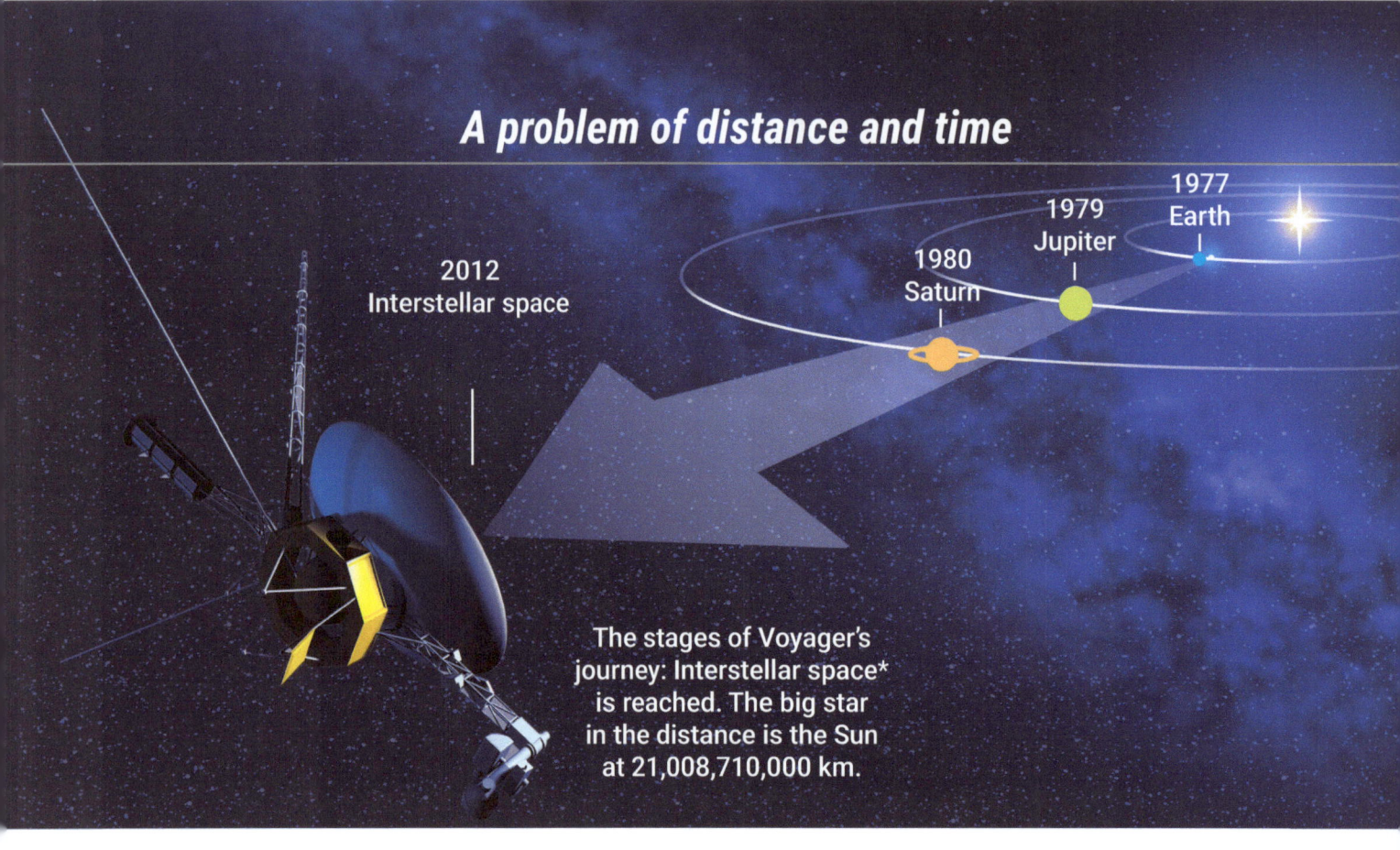

A problem of distance and time

The stages of Voyager's journey: Interstellar space* is reached. The big star in the distance is the Sun at 21,008,710,000 km.

Can we communicate with extraterrestrials?

The Voyager 1 probe, launched in 1977 to study planets, left the solar system in 2012. It is now traveling at 60,000 km/h through interstellar space. Voyager carries a message engraved on a disk, intended to communicate to extraterrestrial intelligence the position of Earth and what its inhabitants achieve in terms of art, science, and culture. Despite its speed, it will take 40,000 years to reach another star. One technical possibility for contact is to use the large antennas of radio telescopes* to send signals at the speed of light, which is 1,080,000,000 km/h. Even at this speed, the message will need 4 years to reach Proxima Centauri, the closest star to the Sun. If there, on an exoplanet, the inhabitants are dinosaurs or sea scorpions, we may have to wait a long time for a response. But if the inhabitants are capable of building radio telescopes and quickly send us a message, we will still have to wait 4 years to receive it.

The message communicates to extraterrestrials the following scientific data:

NUMBERS: a way of counting using only 0 and 1, as in computer science

MOLÉCULES COMPLEXES DE LA VIE: Désoxyribose, Adénine, etc.

DOUBLE HELIX: the diagram illustrates the assembly of carbon molecules into a double helix, the DNA*.

THE RADIO TELESCOPE ITSELF: hoping that for these unknown extraterrestrials this type of technique means something.

ELEMENTS OF LIFE: some sort of pixels representing hydrogen, carbon, nitrogen, and oxygen. Extraterrestrials thus discover that we know the basic elements of life.

HUMANITY: a human of average height.

THE SOLAR SYSTEM: the sun then Mercury, Venus, Earth slightly shifted towards the figure, Jupiter, Saturn, with 3 pixels to indicate their large sizes, then Uranus, Neptune, and Pluto, considered in 1977 as a planet.

Here is the Arecibo radio telescope: in 1975, it transmitted the message above to the Hercules Cluster, a group of stars surrounded by many planets.

25

The best of all worlds

Is Earth a perfect planet?

What makes Earth a paradise for life? Firstly, its orbit stays in the habitable zone all year round. Secondly, its star, the Sun, distributes its warmth peacefully and for a long time. Let's not forget the Moon, which stabilizes the rotation of our planet by orbiting around it, or the atmosphere we breathe, which regulates the temperature of liquid water by controlling its boiling point. Finally, the metallic core, which rotates at the center of the globe, creates a magnetic field that transforms deadly particles from the Sun into auroras borealis. We discover exoplanets almost every day. Some have physical characteristics that also allow for the presence of liquid water, and atoms of carbon, hydrogen, oxygen, and nitrogen can combine into organic molecules, as on Earth. But we still do not know the mechanisms that transform inanimate organic molecules into living cells. The appearance of life on a planet with good physical and chemical conditions is not «automatic». Is Earth the only celestial body that has given birth to five families of living beings? The exploration of other worlds is just beginning. The universe is vast, and the planets to be discovered are more numerous than all the grains of sand on a beach.

SPACE DICTIONARY

Provides the definition of the 34 words marked in the text with an asterisk *

DNA: a molecule present in all living beings

Light-year: the distance traveled by light in one year

Apparent: whose appearance does not correspond to reality (the Sun is much larger than the Moon but as it is also much further away, it appears to be the same size)

Atom: the smallest part of a substance that can combine with another.

Attraction: force that acts like a magnet on any object on the ground or in the air

Aurora borealis: a luminous phenomenon visible in the polar night sky

Nitrogen: an atom whose symbol is N (from the Latin Nitrogenium)

Ice floe: a layer of ice that forms on the surface of a body of water

Carbon: an atom whose symbol is C

Comets: small celestial body made up of a nucleus of ice and dust

Deoxyribose, adenine, etc.: molecules present in all living beings

Elements: groups of atoms or molecules

Interplanetary space: the region of space between planets

Star: celestial body that emits light and heat

Fossil: imprint or visible remains of a very ancient phenomenon.

Hydrogen: an atom whose symbol is H

Interstellar: the space between stars

Tides: deformation of the rocky or liquid globe of a celestial body due to the attraction of another celestial body

Mass: the amount of matter, rock, or gas contained in a celestial body

Meteorite: a natural solid body that crashes into another

Microscope: an optical instrument used to obtain an enlarged image of a very small object

Molecule: a grouping of atoms

Interstellar clouds: accumulations of gas and dust that float in the space between stars

Orbit: the path that an object in space follows when it revolves around another object

Organic: matter made by living beings that always includes carbon.

Oxygen: an atom whose symbol is O

Planet: celestial body orbiting the Sun or another star

Radiation: propagation of energy and motion, often invisible, involving waves, atoms, etc.

Radio telescope: instrument used to capture radio waves emitted by celestial bodies

Satellite: celestial object in orbit around a planet or another object larger than itself

Space probe: unmanned spacecraft launched into space to study various distant celestial objects

Spectroscope: an apparatus that transforms a beam of light into a spectrum (i.e. a sort of rainbow)

Stellar: pertaining to stars

Universe: the entirety of everything that exists

Goodbye!

Iko, Dod et Kani

www.ingramcontent.com/pod-product-compliance
Lightning Source LLC
Chambersburg PA
CBHW051940210526
45473CB00006B/2325